HALLEY'S COMET 1986 ALMANAC

This special publication was produced jointly by the Franklin Institute and the Philadelphia Daily News.

The scientific articles for this section were written by Derrick Pitts, chief observational astronomer of the Franklin Institute's Fels Planetarium. Historical articles were written by Ann Mintz, director of communications and marketing at the Franklin Institute. Classroom activities and parts of the glossary were written by Don Steinberg, director of science education for the School District of Philadelphia. The picture on the front cover was painted by Val Gonzales, assistant director of the Fels Planetarium.

Editor Jerry Carrier
Project Coordinator Lynne Berman
Design Director Joe Curcio
Book Designer Emre Kavlakoglu
Circulation Director Stanley Budner
Circulation Consultant Stan Binder
Publisher Norman Goldfind

Published By
THE MIDDLE ATLANTIC PRESS
848 Church Street, P.O. Box 945
Wilmington, Delaware 19899

All Rights Reserved

Copyright © 1986 by Philadelphia Newspapers, Inc.

All rights reserved. No part of this publication may be reproduced in any form or by any electronic or mechanical means including information storage and retrieval systems without permission in writing from the Publisher, except by a reviewer who may quote brief passages in a review.

ISBN 0-912608-38-2
Printed in the United States of America

The Middle Atlantic Press, Inc.
848 Church Street, P.O. Box 945
Wilmington, Delaware 19899

CONTENTS

	Page
How to Find the Comet	4
From Flaming Spear to 'Dirty Snowball'	8
The Sun Served Up Some Cold Leftovers	12
Comet Could Bring A Gift of Knowledge	14
Close Comet Encounters of the Spacecraft Kind	17
Halley Made The Comet His	20
Halley's Orbit Studies Were on Right Track	24
Comet's Journey Through History	26
Here's a Great System System That's Tough to Beat	36
Reading the Skies Is an Ancient Art	40
Comet Crossword	42
Every Day Is A Halley Day	44
What Do They Mean by That? A glossary words	46
Portraits of a Comet	50
A Halley Album	52

How to Find the Comet

Here's what it's all about. You've read about the comet; you've heard about the comet. Now you want to see it.

By **DERRICK PITTS**
The Franklin Institute

Although much of the history of Halley's Comet is fascinating, none of it will be as important as the history made when you see it for the first time.

Due to the positioning of the Earth through this particular passage of Halley's, we in the northern hemisphere will have to work a little to be rewarded with a good view of this most august visitor to our solar system.

First and probably most importantly, you must observe from a dark sky location. Areas in or around big cities will not offer much opportunity, if any, to see the comet. Bright city lights will hamper any attempt to see it, no matter how big your telescope is. Ideally, a treeless, hilltop site with a clear view to the east, south and west will be best.

With patience, the simple star maps on these pages, and a pair of binoculars, you should be able to find the comet from your dark sky location.

As you read this, Comet Halley has already passed behind the Sun and is on its way out of the Solar System. On February 9, it made its closest approach to the Sun, or reached perihelion. At that time, the Earth was on the opposite side of the Sun, so the comet was quite far away from us. As it emerges from behind the Sun in early March, it is an early morning object, appearing low in the southeast sky between the constellations Capricornus and Sagittarius in the hours before dawn. The comet is in the southern sky and traveling south, so comet viewing in the northern hemisphere is marginal at this time. However, because the comet has just brushed by the Sun, it's near its brightest, and worth trying to see. With any luck, even if the nucleus is below the horizon, the comet's tail should be visible.

The best comet viewing sites in the United States are those in the southernmost parts of the country. The farther south you go, the higher in the sky the comet will be. Try Florida, Texas, Arizona, California or Hawaii, if you can. But wherever you may be, you can expect some interference from the Moon between March 21 and April 2.

At this time, the comet is at its brightest. Its close passage to the Sun has developed the comet's tail to its fullest extent, and the head glows brightly enough to be easily seen without any optical aid. The comet is moving away from the Sun, but it is approaching the Earth. It is closest to the Earth on April 10, and will seem to be moving quickly across the sky.

But for observers in the northern hemisphere, the comet will be low in the sky. North of 43° north latitude, the nucleus and coma will be below the horizon and lost to sight, although as much as 25° of the long, glowing tail may be visible even that far north. South of 35° north latitude, the comet will be low in the southern sky, traveling between the tail of Scorpius and the northern part of Centaurus.

Moving along the eastern horizon towards the south, Comet Halley

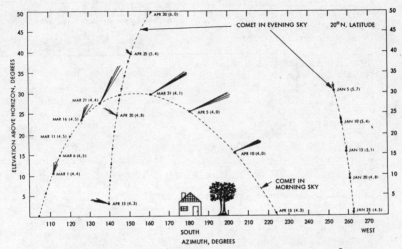

Comet Halley as observed in 1986 by an observer located at 20° north latitude. The comet positions are given for the beginning of morning twilight or the end of evening twilight. Approximate visual magnitudes are given in parentheses following dates.
(From Comet Halley Handbook, Courtesy D. K. Yeomans.)

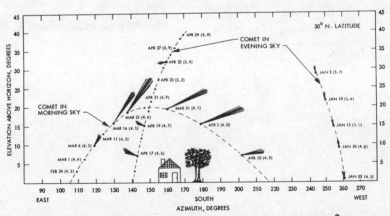

Comet Halley as observed in 1986 by an observer located at 30° north latitude. The comet positions are given for the beginning of morning twilight or the end of evening twilight. Approximate visual magnitudes are given in parentheses following dates.
(From Comet Halley Handbook, Courtesy D. K. Yeomans.)

rises earlier each morning. From April 2 to April 7, you can see it best in the hours before dawn. On the 8th, it will be highest in the sky at 3 am. It will advance about a half hour per night. This means that on April 10, you should look for it at 2 am; on the 12th, at 1 am, etc. After the 13th, although the comet will again be an evening object, you can expect some interference from the waxing moon.

During this spring viewing season, you should be able to find the comet without any optical aids.

After the middle of April, the comet begins to fade rapidly as it moves out of the inner Solar System. It is shrinking and fading as it begins its 38 year journey back to the distant outer reaches of the Solar System, beyond the orbit of Neptune. The comet will be visible in the evening sky, highest and most visible soon after full darkness. By the end of April, you'll need binoculars to spot it. But since the Moon will not interfere with your comet viewing between April 26 and May 10, comet watching should still be possible, with the comet visible in the skies south of Leo and west of Corvus.

Your last chance for comet watching is between May 25 and June 9. By this time, you'll need a telescope to see it. The next opportunity to view Halley's Comet from this part of the Solar System will not be until its next visit, in 2061 A.D.

Halley Hot Line to Be Offered

Associated Press

WASHINGTON — Overwhelmed by the demand for information on Halley's Comet, the U.S. Naval Observatory will experiment with a high-volume comet hot line starting this month.

More than 20,000 calls have come in since Sept. 3, keeping the single line offering recorded Halley information busy nearly 24 hours a day, with an unknown number of callers turned away by a busy signal.

The observatory and the American Astronomical Society will inaugurate a new special line starting at noon, Dec. 15, on an experimental basis. The Halley Hot Line number will be 900-410-8766.

Calls to the new hotline will cost 50 cents for the first minute and 35 cents for each additional minute. The hot line will remain in service until April 15.

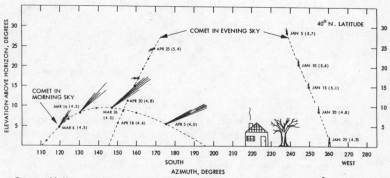

Comet Halley as observed in 1986 by an observer located at 40° north latitude.

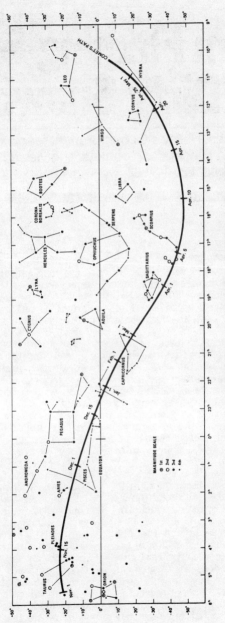

Path of Comet Halley between November 1985 and May 1986.

From Flaming Spear To 'Dirty Snowball'

For centuries, Halley's Comet brought fear with it. People didn't understand comets, and viewed them as omens of disaster. In 1985, we know better.

By **ANN MINTZ**
The Franklin Institute

What is a comet? Where do comets come from? Where do they go? Where is Halley's Comet now? How can I see it for myself? What is all the fuss about anyway?

Halley's Comet is back in our neighborhood for the first time since 1910, and the eyes of the world are on the skies. This is nothing new. Halley's Comet has been visiting our part of the solar system every 75 years for at least 2,000 years, probably much longer. And every time, people noticed, people wondered, and for centuries, people were afraid.

Why be afraid of a comet? It's just one more bright, beautiful sight in the night skies — not that different from the moon or the stars and planets, right?

Wrong. The moon and sun, the stars and planets, are always in the sky. They change slowly through the seasons; the moon goes through its cycle from new moon to full moon every month. They are familiar sights, predictable and certain.

Comets are something completely different. They appear out of nowhere, moving much, much faster than anything else in the sky. They have long, glowing tails, and sometimes a comet's tail can stretch over much of the night sky. A really bright comet can even be seen in broad daylight.

Well, what's so scary about that?

For thousands of years, people feared anything unknown. The world was a mysterious and dangerous place. Anything out of the ordinary was something to fear and avoid.

People believed in omens. They thought unusual events in nature were sure signs of trouble. (For some reason, no one ever thought that omens could be signs of good luck.) Comets were unusual. Comets were bad news. That's all there was to it.

No one knew what a comet was, of course. There's really nothing dangerous about them. They're "dirty snowballs" — frozen gases and water, mixed with tiny bits of rock and grit. Moving through our solar system, they follow the same natural laws as the planets and their moons.

But in the long years before modern science, comets were seen as fiery swords in the hands of angry angels, the cut-off heads of heroes or

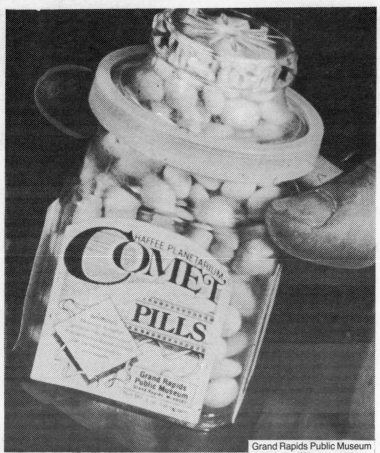

These Comet Pills are a joke; in 1910, people took them seriously

demons, burning spears, stars that had somehow grown hair. In fact, the very word *comet* means "hairy star."

Over the years, comets have been blamed for all sorts of disasters, from the fall of Jerusalem in 66 A.D. to the fall of Quebec in 1759, from Attila the Hun attacking Europe in 451 to Genghis Khan attacking everyone in 1222. Plagues, floods, wars and fires all have been blamed on comets.

Did comets have anything to do with any of these disasters? No, but no one believed that. And all sorts of disasters occurred when there wasn't a comet in sight. No one thought about that either.

Even as late as 1910, many people thought Halley's Comet would bring trouble. Some of the fear was superstitious; some people even thought the world would end.

Others had a more scientific reason to be afraid, or so they thought. Scientists had found tiny traces of cyanogen, a poisonous gas, in the comet's tail. The age-old fear of comets had a brand-new, modern explanation. The Earth would pass through the comet's tail, and every living creature would perish from the poison!

The earth did pass through the comet's tail. But there were such tiny traces of the gas that no one even noticed it. No one, that is, except for the sharp operators who made a fortune selling Comet Pills to protect the people foolish enough to buy them.

This year's visit of Halley's Comet is different from the others. Very few people expect the end of the world or other disasters. People are selling Comet Pills, but it's nothing serious. Just another gimmick, along with Halley's Comet stuffed dolls, glow-in-the-dark T-shirts and comet-welcoming balloons.

This time around, we're being sillier and more serious at the same time. We are asking serious scientific questions, and we'll learn more from the comet than ever before. At the same time, we're greeting the comet with a media blitz, with animated cartoon characters and novelty items. But for the first time in thousands of years, we aren't afraid of this cosmic visitor.

ARE YOU PREPARED TO VIEW

HALLEY'S COMET

WHICH IS NOW RAPIDLY APPROACHING THE EARTH?

AITCHISON & CO.'S SPECIAL "TARGET"
A PORTABLE TELESCOPE WITH PANCRATIC EYEPIECE, gives a magnifying power of 25 30 35 40 diameters, has a 2¼-inch Object-Glass, Leather-covered Body, Caps and Sling. Closes to 11 inches.
PRICE £4 4s. CARRIAGE FREE.
IT IS ALSO A FIRST-RATE TELESCOPE FOR GENERAL USE.

The "UNIVERSAL" Astronomical Telescope
ON TRIPOD TABLE STAND WITH 3-INCH OBJECTIVE AND TWO EYEPIECES. Price Complete in Polished Pine Case. **£5 10s**
Carriage Extra.
Write for Price List of Telescopes and Prism Binoculars.

AITCHISON & Co., Opticians to H.M. and U.S.A. Govts.
428, Strand, London, W.C., and Branches.
Yorkshire Branch: 49, Bond Street, Leeds.

Franklin Institute

Franklin Institute

THE CRAZE OF 1910

Halley's Comet was as big a craze in 1910 as it is now. Telescope makers cashed in on the comet with ads like the one at left. Even political cartoonists used Halley's Comet as material. At right, President William Howard Taft is seen as haunted by his predecessor, Teddy Roosevelt, as he sees Roosevelt's famous glasses and toothy grin in the comet.

The Sun Served Up Some Cold Leftovers

Comets, which scientists now liken to 'dirty snowballs,' are made up of debris that remained when the sun was formed by a giant nuclear explosion 5 billion years ago.

By **DERRICK PITTS**
The Franklin Institute

About 5 billion years ago, our solar system was not much more than a cosmic cauldron heated by the fires of gravitational collapse.

Within an extraordinary cast of chemical materials, the element hydrogen stood out as the most abundant in the cauldron. As gravity pulled it inward, it created some very unstable conditions.

If hydrogen nuclei are heated enough under truly cosmic pressures, they will fuse to produce heavier elements like helium.

But for lighter hydrogen nuclei to become heavier helium atoms, a small loss of mass must occur, and that releases an incredible amount of energy. This hydrogen fusion process is essentially what happens when a hydrogen bomb explodes. This explosion uses only tiny amounts of fuel to produce the most powerful blasts known on earth.

Our sun is a hydrogen-fusion, chain-reaction explosion controlled only by its own mass and its seemingly limitless supply of hydrogen gas. The combined instability of our cosmic cauldron and gravitational collapse produced the temperatures and pressures necessary to begin the thermonuclear process that formed the sun 5 billion years ago.

As only hydrogen was used to start the nuclear furnace of our solar system, the leftover elements and chemical constituents became the building blocks for all the planets and moons we see in our nighttime skies.

Occasionally we see remnants of our cosmic beginnings in the form of comets. They originated as water and frozen gases that moved away from the heat of the young sun to form a cloud at the outermost reaches of the solar system.

This cloud is called the "Oort Cloud," for Jan Oort of the Leiden Observatory in the Netherlands, who first developed the theory that the cloud existed. It could contain as many as 10 trillion proto-comets.

As frozen balls of ice, most comets have spent their billion-year existence in the Oort Cloud some 100,000 times farther from the sun than Earth is. Locked in gravitational orbit around the sun, these proto-comets are *never* seen.

Those that we see are the few that are disturbed in just the right way by a passing star or a large planet to cause them to fall toward the center of the solar system.

For hundreds of years, earth dwellers considered comets to be fiery, hot objects moving through the skies. It was only within recent history that scientists determined that comets are actually extremely cold iceballs, moving in space where the temperature is extremely low, actually reaching minus-459 degrees Fahrenheit, or absolute zero.

Modern spectroscopic analysis of comet heads and tails shows them to be composed of such common elements as carbon, hydrogen, nitrogen and oxygen.

Some of the more mundane combinations of the atoms in comets are methane, ammonia, carbon dioxide and carbon monoxide. In the exotic class are such substances as hydrogen cyanide and the deadly cyanogen gas, the presence of which frightened many people in 1910, the last time Halley's Comet appeared.

During almost all of their journey, comets remain undetected as dark, "dirty snowballs," a term coined by Dr. Fred Whipple of Harvard University in 1951.

They become active and visible only when they approach the sun. Coupled with the understanding of cometary orbits given to us by Edmund Halley, we have learned that, during most of their orbits, comets are dormant.

As a comet approaches the sun, solar radiation warms the frozen gases of the icy nucleus. The warming action converts the ice of the nucleus directly into gases, which surround the nucleus in an envelope called a *coma*.

Development of a *tail* depends on how close the comet comes to the sun. The closer its approach, the greater the amount of material vaporized off the nucleus and into the coma. While comet nuclei range between 1 mile and 40 miles in diameter, the comas can range between 10,000 and 100,000 miles in diameter.

When the comet passes near the sun, the gases of the coma are usually pushed out into a tail extending 100 million miles or more in some cases. Many times, comets don't get close enough to the sun to develop a tail at all.

The major reason why we see comets is because of the sun's light, which is reflected from the dust particles in the coma and dust tail.

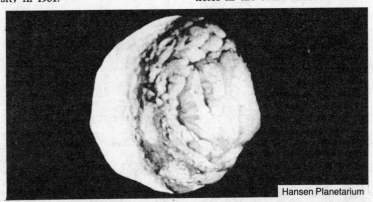

Hansen Planetarium

This drawing shows what scientists believe the "dirty snowball" nucleus of a comet looks like

Comet Could Bring A Gift of Knowledge

Scientists are preparing for some up-close study of Halley's Comet, hoping it will provide some answers not only about comets but other space mysteries, as well.

By **DERRICK PITTS**
The Franklin Institute

As long as comets have existed, we've wondered exactly what they are. With this, the 30th recorded passage of Halley's Comet, we stand to learn quite a bit about comets.

At the last passage of Halley's Comet, prominent astronomers thought comets were "flying gravel banks," loose swarms of separate particles accompanied by dust and gas.

Now we are expecting a "dirty snowball," whose close approach to the sun will melt some of the snowball to become the tail.

We have the ability to send spacecraft out to meet this celebrated visitor, and in doing so scientists hope to answer some of the many questions they have about comets:

■ In some photographs of comets, long sunward spikes called anti-tails are seen. Are we correct in assuming that they are a trick of light?
■ How valid is the theory that a comet's nucleus spins on an axis, exerting a force that alters its orbit?
■ Are the unusual structures that sometimes show up in comets' tails the result of the nucleus' conversion from solid to gas by the sun's heat?
■ The nucleus itself is believed to have a particular structure and deterioration rate that scientists would like to confirm. This would help us determine the life span of various kinds of comets.
■ How much activity goes on inside a comet's tail?
■ Comets are believed to be frozen balls of gas with particles of dust and rock of unknown origin embedded

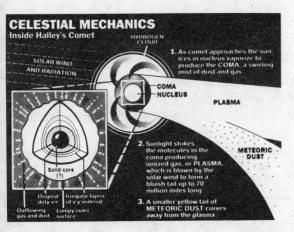

within them. We think they were produced during the beginning of our solar system. If this is true, they may be carrying information about the earliest conditions of our solar system locked within their frozen gases, dust and rocky material.

■ We believe that comets come from the Oort Cloud, which encirles the solar system at a considerable distance from the sun. Scientists hope a close look at the comet can tell us something about the Oort Cloud.

■ As periodic comets make round-trip journeys through the solar system, many scientists are asking if comets pick up material as they move among the planets. Do they recharge themselves as they travel through the dormant parts of their orbits, or do they lose material as they near the sun, eventually dwindling down to nothing?

Comets also provide information indirectly about our sun through our observations of the coma and tail activity. Because the solar wind plays a big part in the development of comet tails, solar physicists can monitor the action of the solar wind by observing the tails.

There are also many questions to be answered about the effect of the gravitational fields of Jupiter and Saturn. These large bodies could substantially alter a comet's path and send it on a different course.

One of the biggest and newest questions about comets is what role a comet may have played in the sudden extinction of dinosaurs some 100 million years ago. Did a comet raise a cloud of cosmic dust that led to the worldwide destruction of dinosaurs, or was some other agent responsible for this hard-to-explain event?

With the close passage of several satellites to Halley's Comet we will probably be able to answer some of these questions and raise many new questions, as well.

A comet has three distinct parts. At its heart is the nucleus, a frozen mass of gas, dirt and rock. Surrounding the nucleus is a cloud called the coma, formed when the sun's heat changes part of the frozen nucleus into gas. The spectacular tail has two parts, the bluish plasma (or gas tail) and the yellowish dust tail. The box at the left of this graphic shows that the nucleus itself has several layers.

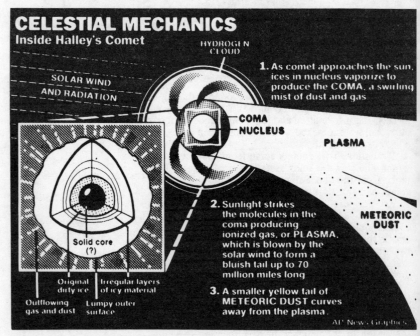

Close Comet Encounters Of the Spacecraft Kind

Scientists got their closest look yet at a comet when they redirected a research satellite into the path of an approaching comet in September. What they learned was fascinating.

With the long-expected passage of Comet Halley, scientists have been watching closely for several years in an attempt to confirm many of their theories about comets.

Using the largest telescopes available, astronomers have been able to identify the comet out between the orbits of Jupiter and Saturn, more than three years from perihelion, its closest approach to the sun. They've even been able to watch the nucleus "turn on" under the influence of solar radiation.

Much of our knowledge about comets has come recently through the use of a satellite previously used to study the effects of the sun's radiation on Earth's magnetic field.

Once known as the International Solar Envelope Experiment, Project Ice was redirected in 1981 to intersect the path of the approaching Comet Giacobini-Zinner in early September 1985.

An encounter of this sort had never taken place before, so scientists were understandably concerned about the meeting, which was to bring the satellite to within 5,000 miles of the comet's nucleus.

Based on what they believed at the time about what would happen when another object approached a comet's nucleus, project leaders predicted a 50-50 chance of serious damage to the spacecraft through dust particle bombardment or voltage reduction caused by dust accumulation on the spacecraft's solar panels.

At 7:30 a.m. on Sept. 11, Project Ice scientists were elated to find that the spacecraft had safely negotiated the comet's tail with little impact by either dust particles or magnetic fields.

Extremely sensitive instruments on board the space laboratory discovered three things about the comet:

■ The coma contains an unusually small number of dust grains.

■ Water ions derived from the nucleus are extremely active.

■ The charged particles within the plasma have very low kinetic energy, the energy generated by motion.

These discoveries alone will force researchers to reconsider several theories of comet behavior.

—DERRICK PITTS

WELCOMING COMMITTEE

Scientists all over the world are excited about this year's visit of Halley's Comet. For the first time, we have the technology to get a really close look — with satellites. The European Space Agency will send the Giotto (above) into the comet's path, while the United States, the Soviet Union and several other countries will observe the comet with the Soviet-launched Vega (below).

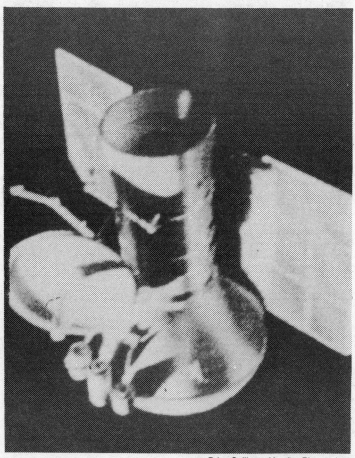

Brian Sullivan–Hayden Planetarium

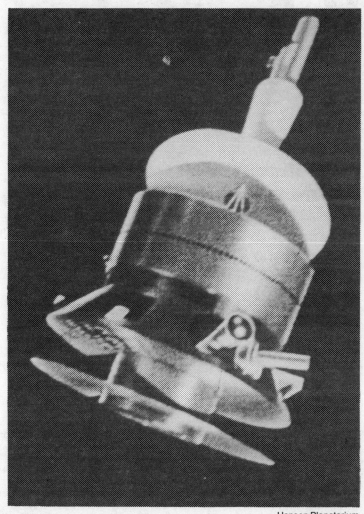

Hansen Planetarium

Halley Made The Comet His

Edmund Halley didn't "discover" the comet that bears his name, but he was the first astronomer to predict when it would be seen again.

By **CAROLEEN VAUGHN** and **ANN MINTZ**
The Franklin Institute

Edmund Halley was born in England in 1656, when many people still believed in witchcraft, omens and spells. One person in five died of the plague — the "Black Death" — and people feared the unknown. Anything unusual or unexplained could cause panic.

But the unexplained did not frighten Halley. It sparked his curiosity and inspired a brilliant career. He was a man of letters as well as science, and his inquiring mind led him to a whole galaxy of discoveries.

Halley was no ordinary man. He had influential friends all over the world — from Czar Peter the Great of Russia to Isaac Newton, the greatest scientist of his day. Halley swore like a sea dog, and was known for his fondness for brandy. But we remember him today because he was the first astronomer to correctly chart the course of a comet and predict accurately when it would next be seen. The comet whose return he so successfully predicted carries his name today.

As the son of a wealthy soapmaker, Edmund Halley was able to attend Queen's College, Oxford. Shyness was never one of his problems. He exchanged letters with the most important scientists of his day, including John Flamsteed, astronomer royal of England, while still a schoolboy. And he published his first serious scientific work at the age of 19, while still a student at Oxford.

By the time Halley left Oxford in 1676, he already had published three scientific papers. He wanted to accomplish something useful. "I would very willingly do something to serve my generation," he said.

With the help of a generous allowance from his father of 300 pounds a year, he found something useful to do.

Until this time, astronomers had studied only the familiar stars of the North. The stars of the Southern Hemisphere were a mystery, invisible from Europe, known only by legend, rumor and sailors' stories.

Halley decided to make scientific observations of the southern stars, and he chose the lonely South Atlantic island of St. Helena as the location for the first observatory ever in the southern half of the world. King Charles II of England granted him

Edmund Halley had an illustrious career as England's premier astronomer

free passage to St. Helena and free housing when he arrived, and Halley set sail.

Unfortunately, the expedition was not a complete success. St. Helena's climate is not very good for astronomical observation. When Halley returned to England in 1678, he had catalogued only 341 new stars. Still, this was 341 more than had been known before, and Halley's reputation as an astronomer was now assured. He was elected to the Royal Society, the most important scientific organization in England, and was awarded a master's degree from Oxford even though he had never really met the requirements.

He was only 22, but he was the darling of the astronomers, the Royal Society, and of England's sea captains as well. His catalog of stars made it possible to sail more confidently in southern waters. This was important to England, since new trade routes to the south were being established. Accurate information was needed, and much appreciated.

After his return from St. Helena, Halley traveled widely throughout Europe. His interest in comets was probably inspired by a visit to the German astronomer Johannes Hevelius in 1679. In 1682, while visiting Paris, he viewed the comet that would later bear his name.

Back in England, Edmund Halley became friends with the great Isaac Newton, a friendship that would last a lifetime. Halley was able to help Newton publish his *Principia*, one of Newton's most important books. Sadly, it was a personal tragedy that made this possible. Halley's father had been murdered in 1684. The money he inherited was enough to support him and his studies for the rest of his life.

Newton helped Halley with his studies of comets. What comets had been observed over the centuries? What records had been kept? What paths did they take across the skies? Did the principles of gravitation that Newton had discovered apply to comets as well as to apples falling from trees and planets moving around the sun?

By the time Halley had observed dozens of comets himself and studied the records of many more, he began to see a pattern. The comet of 1682, which he had seen himself, followed the same path across the sky as the comets of 1456, 1531 and 1607 — every 75-76 years. Could they possibly all be the same comet?

Until Halley's time, people thought comets were rogues of the sky, traveling through our solar system on a one-way path and never seen again. Halley's great insight was that they were true members of the solar system, moving along incredibly long orbits. Based on this insight, he made a prediction: The comet of 1682, 1607, 1531 and 1456 would return in 1758, 76 years later.

Halley lived a long, productive life. He died in 1742 at the age of 85. He accomplished a great deal in many fields, and ended his life as astronomer royal of England. Unfortunately, he did not live to see his prediction come true.

But when the comet returned in 1758, right on schedule, it was given Halley's name. He had not asked for this recognition, wishing only that it would be remembered that the first successful prediction of a comet's return had been made by an Englishman. But Halley's Comet it became, and Halley's Comet it still is today.

Isaac Newton: his studies made it all possible

Halley's Orbit Studies Were on Right Track

Edmund Halley made his prediction of when the comet would return by using new discoveries in math and science to form a theory about the shape of the comet's orbit.

By **DERRICK PITTS**
The Franklin Institute

Halley's Comet was the first comet whose return was accurately predicted. Appropriately, it is named for the person who computed its orbit, determined its period (the time between visits), and predicted its date of return to perihelion (the point in its orbit when a comet is closest to the sun).

Edmund Halley began with the notion that a number of comets sighted over a period of 1,500 years were in fact the same comet. A stumbling block in his way toward fully understanding how these all could be the same comet was that he didn't know the orbital mechanics of these celestial visitors. Were they one-time visitors, or did they keep coming back? If they were repeat visitors, what orbital shape allowed them to return?

Although Halley's education was fairly complete in the known mathematics and physics of the 17th century, calculus, a new branch of mathematics which would help him considerably, was being developed at the time by a contemporary of his, Isaac Newton. Newton also was trying to devise a method of understanding why bodies attract each other.

It was Newton's work in the fields of gravitation and calculus that gave Halley the tools he needed to determine the shape of cometic orbits.

Were these orbits parabolic, hyperbolic or elliptical? With parabolic or hyperbolic orbits, both open-ended curves, comets could easily be one-time visitors, passing near Earth as they flew through space. But if they were periodic visitors, returning at regular and predictable intervals, what force attracted them? Applying Newton's new law of gravity to the series of comets he suspected were the same comet, Halley found his elliptical (egg-shaped) orbit theory fit the observed passages accurately.

Using Newton's newly published work, Halley showed that the sun was both the attracting force and the focus of the elliptical orbits along which some comets traveled. Using this theory, Halley went on to predict that the comet of 1682 would return in 1758.

Halley made his prediction more than 30 years before the comet's expected return to perihelion. Unfortunately, he didn't live long enough to see if his prediction was correct; he died in 1742. The comet is named for Halley, though, because on Christmas Day 1758, Halley's Comet was sighted on its way toward perihelion, due in March 1759.

We now know that comets do indeed have three types of orbits — hyperbolic, parabolic and elliptical. Those whose orbits are hyperbolic or parabolic are one-time visitors. In fact, some of the most spectacular comets ever seen have been ephemeral (non-periodic) comets.

Of the comets with elliptical orbits, Halley's is the brightest short-period comet known. Comet Encke, the shortest short-period comet, reaches perihelion every 3.3 years, while most long-period comets average 40,000 years between appearances.

Franklin Institute

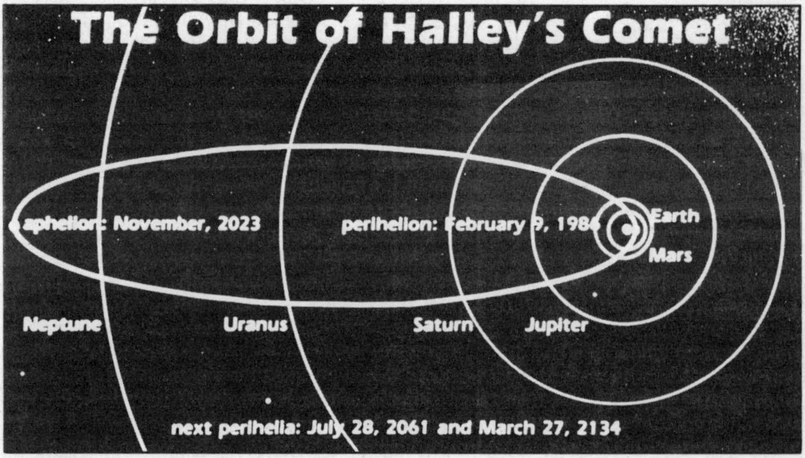

The Orbit of Halley's Comet

aphelion: November, 2023
perihelion: February 9, 1986
Earth
Mars
Neptune Uranus Saturn Jupiter

next perihelia: July 28, 2061 and March 27, 2134

This chart shows the elliptical shape of Halley's Comet; it was Halley's theory on the shape of the orbit that led to his famous prediction of its return

Comet's Journey Through History

A look at how civilization progressed from the first time the appearance of Halley's Comet was recorded until the comet's most recent visit.

Halley's Comet has been visiting us regularly for many centuries. Some record of its appearance has been made for each visit since 240 B.C., but it's likely that the comet has been coming around since long before then.

While the comet usually returns every 75 years, the time between visits can be longer or shorter. Minor variations in its orbit can add or subtract a few years.

A description of life on Earth during each appearance of Halley's Comet is a chronicle of the development of humankind itself. We have progressed slowly but surely from visit to visit:

240-239 B.C.

In China, the first written record is made of a sighting of Halley's Comet, recorded in the Chronicles of Shih Chi and Wen Hsien Thung Kao.

■ Leap Year is introduced into the Egyptian calendar.

■ The Greek scientist Eratosthenes (276-194 B.C.) is the first to suggest that Earth moves around the Sun instead of the Sun moving around the Earth. He also makes the first accurate estimate of the Earth's circumference — the distance around the Earth at the equator.

164 B.C.

It is not clear whether this appearance of Halley's Comet was recorded. In 162 B.C., a "star with a pointed tail" was reported in China. This may have been Halley's Comet.

86 B.C.

Halley's Comet again is reported in China in the Chronicles of Thung Chien Kaung Mu.
- The Greek physician Asclepiades practices natural medicine in Rome.
- After civil war, the Roman Empire triumphs in Italy.

11 B.C.

The great Roman general Marcus Vipsanius Agrippa dies in Rome. Halley's Comet can be seen in Rome when he dies, and the comet is called an omen of his death. This is the first time the comet is called an omen of disaster. It is also well described in Chinese records.
- Augustus is emperor of Rome, from 30 B.C. to 14 A.D. "He found Rome a city of brick and left it a city of marble."
- The Roman Empire starts to expand, moving into Egypt, Judea and Germany.
- In Rome, streets are paved for the first time.
- The use of gears leads to the first animal-powered water wheel for irrigating fields.
- Rome begins to rule the world.

In 66 A.D., the Emperor Nero began persecuting the early Christians

66 A.D.

The historian Flavius Josephus describes Halley's Comet in his "history of the Jewish War," describing the rebellion of Judea against Rome. The city of Jerusalem is destroyed. Josephus says that "among the warnings a comet ... [whose] tail appears to represent the blade of a sword was seen above the city, a comet that continued the whole year." Comets are confirmed as omens of disaster. Again, the comet is well described in Oriental records.

■ The Gospel of St. Mark is written, and a year later, those of St. John and St. Matthew.
■ The Emperor Nero begins the persecution of early Christians.

141 A.D.

Chinese records report that Halley's Comet was especially bright, with a very long, bluish-white tail. The comet is visible for four weeks.

■ The Roman Empire is at its height, covering much of the known world.
■ Antoninus Pius is emperor of Rome.
■ The earliest known Sanskrit inscriptions are written in India.

218 A.D.

The comet is recorded in both Europe and China. The Greek writer Dio Cassius describes it as a fearful star with a tail extending from west to east. It is also reported in the Chinese astronomical chronicles Hou Han Shu, Thung Chien and Ma Tuan Lin.

■ Roman persecution of the early Christians increases.
■ Roman citizenship is granted to every free-born resident of the Empire.
■ In China, the ancient Han dynasty ends, followed by 400 years of civil strife.

295

Halley's Comet is brilliant and low in the eastern sky, according to contemporary records.

■ The Roman Empire begins to break up under pressure from many wild tribes — among them the Goths, Visigoths, Ostrogoths and Franks.
■ Five simple machines in common use are described by Pappus of Alexandria: the cogwheel, lever, screw, pulley and wedge.

374

Scientists estimate that Halley's Comet passed within 9 million miles of Earth. For a comet, that's a near-miss. It was recorded in China in the Chin Shu chronicle.
■ In Europe, another wild tribe, the Huns, invade. Halley's Comet is blamed.
■ The Roman Empire is officially divided into an East Empire and a West Empire.
■ Roman troops leave Britain.
■ In Europe, books begin to replace scrolls.
■ The first written records of Japanese history are made.

451

Attila, king of the Huns, invades France but is defeated in one of the most terrible battles of history, the Battle of Chalons. A great, long-tailed star was shining during the battle — Halley's Comet, which was blamed by many first for the invasion and then credited with Attila's death.

Attila

■ The city of Venice is founded in Italy by people fleeing the Huns.
■ An Indian astronomer publishes a mathematical work on the powers and roots of numbers.
■ The Roman Empire begins its final collapse.

530

Although this was one of Halley's Comet's least spectacular appearances, it was recorded both in Europe and China.
■ The Byzantine emperor Justinian issues his code of civil laws. He also closes the 1,000-year-old School of Philosophy in Athens.

607-608

In China, the Pei Shih Chronicle describes a tailed star across the heavens, and the Wen Hsien Thung Kao Chronicle also describes Halley's Comet. It was visible for an unusually long time, and came quite close to the Earth.
■ Mohammed, founder of Islam, is living in Arabia.
■ Books are printed in China.
■ Smallpox moves into Europe from India.

684

Although the comet was not spectacular in 684, it was recorded all over the world. A drawing made in 684 appears in the 15th-century Nuremberg Chronicles. The Nuremberg Chronicles note that the comet brought three months of rain, thunder and lightning. People and animals died, grain withered, and a plague followed eclipses of both the sun and the moon. Halley's Comet was also described in both Chinese and Japanese chronicles.

■ In his description of the Comet of 678 (not Halley's), an English churchman, the Venerable Bede, says that comets bring disaster, war, disease and the death of kings.

760

Chinese records describe a "broom star, white in color."

■ Arabic numbers appear in Baghdad.
■ The Turkish Empire is founded.
■ In Central America, the Mayans learn enough astronomy to devise extremely accurate calendars. They build temples to the Sun and the Moon.
■ The Book of Kells, "the most beautiful illuminated manuscript in the world," is written in Ireland.

837

Halley's Comet makes its most spectacular appearance, only 3 million to 4 million miles away from the Earth. It was described all over the world, and its tail filled most of the sky. It was so frightening that Louis I, emperor of France and Germany, built many new churches and monasteries.

■ Vikings from the North roam the civilized world.
■ The word "algebra" is invented in Persia.
■ Arabs invade France and settle in Italy.

912

This appearance was so unspectacular that it was recorded very little. It was mentioned in the Japanese Dainihonski, and perhaps in China.

989

Again, this was not a dramatic appearance. It was recorded by several writers in England and in China.

Halley's Comet is depicted in the Bayeux Tapestry (just to the right of the word "Stella"), which marked the Battle of Hastings

- A few years earlier, Otto I founds the Holy Roman Empire in Germany.
- The number system we use today was introduced into Europe by the Arabs.
- Arab science, art and philosophy reach their greatest heights.

1066

This is probably the most famous appearance in history. It was also one of the most spectacular, with the tail filling much of the sky. It was recorded in China and Korea as well as in Europe.

Halley's Comet was in the sky just before the Norman Conquest of England. William the Conqueror defeated Saxon King Harold Fairhair at the Battle of Hastings. William's wife, Queen Matilda, and her ladies embroider the Bayeux Tapestry to celebrate the victory. It shows the comet in the skies with people gazing at it in awe. The Anglo-Saxon Chronicle records that "then was all over England such a token seen in the heavens such as no man ever saw before. Some men said that it was Cometa the star."

- The astrolabe, an important astronomical tool, is brought to Europe from the East.
- Macbeth, king of Scotland, is defeated. His story is the basis for one of the most famous plays of William Shakespeare.
- In England, Westminster Abbey is consecrated.

1145

Chinese chronicles say that the comet was first seen on April 26, in the morning sky. It had lost its tail, but was still seen as a "guest star" by the Chinese.

- In France, the music of the troubadours is heard for the first time.
- The Second Crusade fails, and many crusaders die in Asia.

1222-1223

Genghis Khan, leader of the Mongol Horde, invades China and moves on to India and Persia. He believed Halley's Comet brought him luck and called it his special star. It brought bad luck to the millions he killed.

- Henry III, the boy king of England, is only

Genghis Khan

9 years old. The citizens of his troubled country blame England's problems on the comet.
- The Mongols invade Russia.
- Cotton is introduced to Spain.
- Genghis Khan dies soon after and his huge empire is divided among his three sons.
- Vienna becomes a city.

1301

Halley's Comet is recorded from Iceland to Japan. The great Florentine painter Giotto uses Halley's Comet to depict the Star of Bethlehem in his fresco, "Adoration of the Magi."
- Edward I of England invades Scotland, starting centuries of war. Temporary truce follows.
- In Germany, medicine is dispensed by apothecaries.
- In England, the king defines a standard yard and a standard acre.

Copernicus

1378

This appearance goes almost unrecorded, though it is mentioned briefly in China.
- The Great Schism begins, and for years, there are two popes, one in Rome and one in France.
- Playing cards are introduced in Germany.
- Robin Hood makes his first appearance in English literature.

1456

Pope Callistus III excommunicates Halley's Comet as fraudulent and un-Christian. He prays for delivery from "the devil, the Turks and the Comet." The Turks were threatening all of Europe, but were defeated in the Battle of Belgrade by Jan Hunyady. Many still believed the comet to be an omen of disaster, even though the Turkish invaders were defeated.

1531

For the first time, Halley's Comet is recorded in Europe before it is described in China. It is also described by Korean astronomers. In Europe, it causes a great wave of superstitious fear. In France, Peter Apian observes that the tail of the comet always points away from the sun.
- The great age of European exploration begins in the New World.
- Henry VIII founds the Church of England and cuts all ties with the Roman Catholic Church.
- The great scientist Nicolaus Copernicus is alive (1473-1543). He makes many important discoveries in his career, and is often called the father of modern astronomy.

1607

The great astronomer Johannes Kepler observes Halley's Comet from Prague. Modern astronomy has begun, and European records are more and more detailed and useful. In Virginia, they call the comet "the Red Knife."
- England is ravaged by the plague; the comet is blamed.
- Jamestown, the first English settlement in the New World, is founded in Virginia.
- Galileo constructs an astronomical telescope.

Henry VIII

1682

It was this appearance of Halley's Comet that was actually observed by Edmund Halley himself. Oddly enough, it was not a spectacular appearance. Another comet which had been seen a few years before was much brighter, but much less famous. It inspired one of the first books against comet superstition.
■ Pennsylvania receives its royal charter.
■ The Louisiana Territory is claimed for France by La Salle.
■ Isaac Newton explains the tides using his theory of gravitation.

1759

Edmund Halley successfully predicted that the comet would return in 1759. This was the first time a comet's return was ever actually predicted, and the year when Halley's Comet received its name. It also was one of the finest appearances in modern times.
■ The first public concert takes place in Philadelphia.
■ Quebec falls, and French power in the New World comes to an end. Halley's Comet is mentioned as the cause.

1835

Astronomers begin to make physical studies of comets. Accurate observations are made throughout Europe.
■ Phineas T. Barnum, founder of Barnum Circus, begins his career as a showman.
■ The New York Herald begins publication.
■ The first negative photograph is taken in England.

1910

The last great appearance of Halley's Comet. It was preceded unexpectedly by "The Great January Comet," which appeared just a few months earlier. For the first time, photography was used to study the comet.
■ Marie Curie publishes her book on radiography.
■ The first expedition explores the deep sea.
■ King Edward VII of England dies. His death is blamed on Halley's Comet. Not that much has changed in 2,000 years!

Barnum

Curie

Here's a Great System That's Tough to Beat

9 planets, 44 moons and millions of asteroids, meteors and comets revolve around the sun, which is the only one of the group to generate its own heat and light.

By **DERRICK PITTS**
The Franklin Institute

Thirty thousand light years from the center of our galaxy — the Milky Way — lies an insignificant yellow star.

Only 5 billion years old, this average-sized, average-brightness nuclear furnace sports a gathering of nine planets, 44 natural satellites (moons), and millions of asteroids, meteors and comets.

Only one of those planets supports life. Along with eight other lifeless bodies, it orbits the star at a rate based on its mass and distance from that star.

The star is the sun, and the planet, of course, is Earth.

The **Sun** is the gravitational and geographic center of our family of planets, and all objects within the vast gravitational reach of our star succumb to its command.

Because it is a star, the sun is very different from everything else in our solar system. The sun produces all of its own light and heat. The other objects, in orbit around it, do not. The sun is a huge sphere of very hot gas — nearly 1 million miles in diameter with a surface temperature of about 11,000 degrees Fahrenheit.

If you strung them like beads on a string, you would need more than a hundred earths to span the sun!

The sun is made up mostly of hydrogen and helium, but almost every other known element can be found in it. It rotates on its axis once every 28 days and produces an extraordinary amount of both safe and deadly radiation. In contrast, its planets are for the most part cold, lifeless, unexciting orbs whirling in the darkness of space.

The four planets closest to the sun enjoy the benefits of warmth and light, qualities denied to the five other planets. **Mercury,** the planet closest to the sun, is the second-smallest in the system. With no atmosphere, Mercury, which is less than half the size of Earth, bakes on the side facing the sun and freezes on the other.

A little farther away is cloud-covered **Venus.** Sometimes referred to as Earth's "sister planet," Venus is similar in size to our own planet. The thick, dense clouds that perpetually enshroud Venus keep infrared radiation from escaping. As a result, temperatures are high enough to cook a normal dinner without an oven in about half the time needed on Earth. The thick clouds also produce an atmospheric pressure many times that of the Earth's gentler blanket.

At an ideal distance from the sun, **Earth** is almost completely covered with an extremely rare commodity —

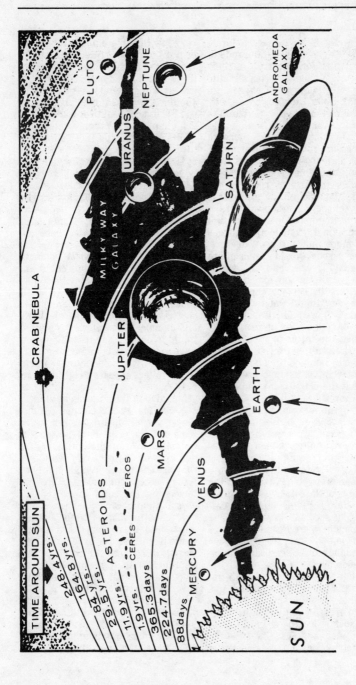

This map of the solar system shows each planet's relative size and distance from the sun, which provides light and heat through a controlled nuclear explosion

liquid water. With rather moderate temperature extremes, thousands of different life forms carry out their sustaining biological processes under a protective layer of oxygen, nitrogen, carbon dioxide, ozone, etc. These life-giving gases and others mixed into our atmosphere shield Earth's inhabitants from otherwise deadly cosmic radiation and meteors.

More than 100 million miles from the sun, the fourth planet, **Mars**, completes the list of rocky inner planets. About 50 million miles from Earth, Mars begins to show the effect of increasing distance from the sun. Its maximum daytime temperature is rarely more than minus-70 degrees. Its thin carbon dioxide atmosphere alternately freezes and thaws depending on the season, expanding or contracting its dry ice caps in the process. Any liquid water that once may have flowed on its surface is now long gone. We once thought Mars had life forms of its own, but unmanned laboratory spacecraft from Earth sampled the Martian soil and determined through metabolic tests that the planet is barren of life as we know it.

The rhythmic spacing of the inner planets is broken by the great distance between Mars and the giant of the system, Jupiter. Spread throughout the intervening distance is a swarm of asteroids in orbit about the sun. Thought to be left over from the beginning of the solar system, the **Asteroid Belt** contains material that never coalesced into a planet as the other members are assumed to have done. Some of the fragments are large enough to be seen in medium-sized telescopes.

Ten times the size of the earth, **Jupiter** is the first of the so-called "gas giants" of the solar system. As Jupiter is composed mainly of lighter elements such as hydrogen, helium, ammonia and methane, its mass is extremely large, but its density is very low. If a big enough tub could be built for it, and filled with water, Jupiter would float! Some scientists think that if it had had a bit more hydrogen when the solar system began, Jupiter too would have become a star. The most beautiful of the planets when seen through a telescope, Jupiter shows many interesting features, such as its great red spot and at least 16 moons, four of which can be seen easily with binoculars.

The wonder of the solar system next in line beyond Jupiter is the ringed planet, **Saturn.** It once was thought to be the only planet with rings, but rings have since been observed around Uranus and Jupiter. The last of the "classical" planets (those which the ancients could see without optical aid), Saturn also has high mass but low density. About 10 times the size of Earth, Saturn is composed of rather light elements and chemicals. Its interior is mainly liquid hydrogen, and it is thought to have a small rock core. Because of its size and high reflectivity, Saturn, like Jupiter, appears brighter than most stars.

Included in the class of gas planets are **Uranus** and **Neptune**. At distances of 1 billion and 2 billion miles from the sun, respectively, these frozen planets orbit the sun every 84 and 165 years, respectively. Although they are composed predominantly of frozen gases, both planets are believed to have solid cores and ice crystal atmospheres. Four times bigger than the planet Earth, Uranus and Neptune are very hard to see without a telescope. In fact, Uranus wasn't discovered until 1781, and Neptune wasn't discovered until 1845. The rings of Uranus were discovered by accident in 1977.

Once known as "Planet X," the system's last planet, **Pluto**, is believed to be half the size of Mercury. At a distance of about 3.5 billion miles from the sun, Pluto receives hardly any radiation. As the outermost planet, it orbits the sun once every 248 years, and the sun would appear 1,000 times dimmer on Pluto than it does on Earth. Although Pluto has no rings, a moon was discovered in 1978. Because of its size and orbital position, Pluto's status as a planet is still in question. Some astronomers believe it could have been a satellite of Neptune at one time, although the discovery of Pluto's own moon, Charon, weakens that idea.

Want to Learn More? Books Are the Way

If you would like to know more about Halley's Comet, here is a suggested list of books you could read, provided by Kit Breckenridge of the Free Library of Philadelphia:

- **Comets and Meteors,** by Isaac Asimov. Follett, 1972.
- **Comets, Meteors and Asteroids,** by Melvin Berger. Putnam, 1981.
- **Comets,** by Franklyn Branley. Crowell, 1984.
- **Halley, Comet 1986,** by Franklyn Branley. Lodestar, 1983.
- **Comets and Meteors,** by George Fichter. Watts, 1982.
- **The Long Journey from Space,** by Seymour Simon. Crown, 1982.

For more general information about astronomy, the science of the stars, the library suggests the following books:

- **How Did We Find Out About the Universe?** by Isaac Asimov. Walker, 1983.
- **Sun Dogs and Shooting Stars: A Sky Watcher's Calendar,** by Franklyn Branley. Houghton, 1980.
- **Astronomy,** by Heather Couper. Watts, 1983.
- **National Geographic Picture Atlas of Our Universe,** by Roy Gallant. National Geographic Society, 1980.
- **The Planets: Exploring the Solar System,** by Roy Gallant. Four Winds, 1982.
- **Sun Up, Sun Down,** by Gail Gibbons. Harcourt, 1983.
- **Cosmic Countdown: What Astronomers Have Learned About the Life of the Universe,** by Francine Jacobs. M. Evans, 1983.
- **Journey to the Planets,** by Patricia Lanker. Crown, 1982.
- **The Young Astronomer's Handbook,** by Ian Ridpath. Arco, 1984.

Reading the Skies Is an Ancient Art

With modern city lighting, the heavenly bodies that light up the night sky are not as familiar to us as they were to our ancestors. But you can still spot planets and constellations with a little training and practice. Just use the horizon.

By **DERRICK PITTS**
The Franklin Institute

For thousands of years the night sky was a constant friend, companion and directional guide.

But with the coming of the industrial age and the brightening of our cities with street lights, our ability to see the sky from large urban areas has diminished. Many of us have forgotten the stories of the constellations and no longer recall which planets are visible in the morning or evening sky.

The coming of Halley's Comet gives us a wonderful opportunity to reconnect ourselves with our sky companions of long ago. Although Halley's will require some familiarization with the constellations of the Zodiac during the late fall, by early winter the sunset horizon will be our most effective guide to finding Halley's Comet and other objects.

About 1½ hours after sunset around Dec. 15, Halley's will be about 60 degrees above the southern horizon, moving progressively westward each night. This time of evening is the perfect time for planet spotting.

In the time between sunset and full darkness, called evening twilight, the first bright objects always seen are planets. In fact, the largest planet of our solar system, Jupiter, has been visible in just this manner for the past three months. If the sky is clear this evening, go out for just five minutes during evening twilight. Scan the southwestern section of the sky until you find a rather large, bright, non-moving object. Does it twinkle? If not, you've found Jupiter.

The best way to planet-spot is to go out just after sunset every evening when the sky is clear for no more than 10 minutes, just to scan whatever horizons are clear to you where you live. To reinforce your discovery, see how far the planet you've spotted is from some familiar object on your skyline.

Now try locating the planet on the next clear evening by scanning the horizon, or better yet finding it from your familiar landmark on the horizon. If you observe at the same time each evening, the planets will move only very small amounts each day. You should begin to notice that they have actually moved in about 1½ weeks.

If you begin your observation early this month, notice that sunset comes a little bit earlier each day. After Dec. 21, though, sunset will start to come later and later, as each day gets longer.

As the sky darkens during evening twilight, you'll see only the brightest stars first. As evening deepens, more and more stars will begin to show themselves until the sky is completely dark and relatively faint stars become visible.

This process takes about 45 minutes.

While you're watching the sky darken, look for the moon, too. During its waxing crescent phases, it makes its first appearances out of the west just after sunset. Under a clear sky, we see one of the finest pictures nature has to offer: a very thin, usually large crescent moon on the background of a colorful sunset, occasionally with bright planets nearby.

A good way to learn about constellations is to look for the brightest stars of prominent constellations right at the end of the evening twilight. If done at this time, *only* the brightest stars are visible. This makes it easier to pick out the general shape of some star groups.

If for some reason you're up in the morning before sunrise, there's morning twilight. It lasts about 45 minutes and is that time period before sunrise.

Morning twilight is actually a better time for planet-spotting than evening twilight because the sky is already dark, and the atmosphere tends to be much clearer. Looking toward the eastern horizon now, we would be looking for morning planets.

In December of this year, the morning sky offers an interesting grouping of three planets,

Mercury, Venus and Saturn, located very low in the southeast. They should be easier to spot on Dec. 10, when the moon joins this triad.

Mars is well up in the southeast during morning twilight, and its reddish glint should be easy to spot. Just as you did with the evening twilight observations, don't spend more than 10-15 minutes per session.

After you become accustomed to this type of observing, you'll find it takes very little time to appreciate the motion of the planets around the sun and you'll be able to recognize the bright planets by name just as you'd know the streets of your neighborhood.

Comet Crossword

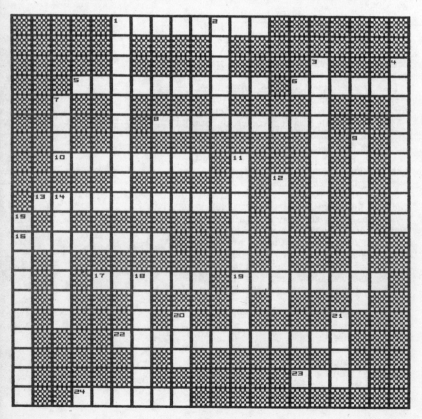

Clues

Across

1. The point at which an object is farthest from the sun
5. What tiny particles do when struck by light
6. A celestial cloud of gas and dust
8. A particle that survives its collision with a planet or moon
10. An angular measurement of position from the horizon upward
13. Counterglow
16. The apparent path of the sun through the starfield
17. The point in the sky that is directly overhead
19. A planet-like body too small to be called a planet
22. The process by which a gas emits light
23. A point along the orbit of an object where it crosses a reference plane
24. A "gas" of positive and negative ions

Down

1. The period of time that a celestial object is visible from Earth
2. The process by which an atom gains or loses an electron
3. A natural particle in space before it enters Earth's atmosphere
4. A measure of brightness in astronomy
7. The cloud of gas and dust surrounding the nucleus of a comet
9. A listing that gives a celestial object's positions at various times
11. Particles in the dust tail that are subjected to equal force
12. An angular measurement of position along the horizon
14. The geometric figure formed by the orbits of planets, asteroids and comets
15. The point in an object's orbit when it is closest to the sun
18. The central mass of a comet
20. An atom that becomes electrically charged by gaining or losing one or more electrons
21. The nucleus and coma of a comet

Every Day Is A Halley Day

This year's appearance by Halley's Comet provides an opportunity for some imaginative learning. Here are some ways your children and their classmates might mark the event.

Oral History

There are people alive today who were living in 1910 when Comet Halley last was visible on Earth. Seek out people in your community who can recall the last appearance of the comet. Retirement communities would be a good source of information.

Tape record your conversations with these people and then report on these interviews. Ask about what life was like in 1910. What did people use for transportation? What appliances did they have? What did they do for movies? What were their schools like? What sort of work did their parents do at that time? Was there a lot of excitement over the comet?

Return of the Comet

Comet Halley will return in the year 2061. The world as we know it will be much different from today. Write an essay describing what you think the world will be like in 2061. Start at the time someone gets up and take that person through the day until he or she goes to sleep that night.

What sort of jobs will people have? How will they get around? Will there be cars? What will the schools of 2061 be like? How will the cities be structured? Choose one area and describe what you believe society will be like in 2061.

Motion of Earth & Sun

The motion of the Earth around the sun is similar to the motion of the comet as it revolves around the sun, except it takes 75 years for the comet to make one revolution, while the Earth completes its journey in one year.

Establish some object to simulate the sun. With your fellow students, walk in an elliptical path around the object. This represents the passage of one year.

Now make one complete rotation in place. This represents the earth rotating on its axis. Each rotation equals one day.

Now rotate and slowly move around the object. This is how the earth rotates, and revolves around the sun.

Halley in the News

In the next several months you'll see increased news coverage of Halley's Comet. Clip articles from the newspaper and display them on a bulletin board.

Shadow Play

The sun rises in the east and sets in the west while appearing to move across the southern sky. A simple sundial can help to illustrate this appearance of motion by the sun that is caused by the earth's rotation west to east.

Take a thin object (maybe a Popsicle stick) and place it upright in a classroom window that faces south. Mark the shadow's location at regular intervals (one hour). Do this for several days, and then use marks to predict the shadow's location for a particular time of day.

Go into the schoolyard and mark the length of your shadow. Repeat this activity once a week over the course of the school year. When will the shadow be the shortest? When will it be the longest? What direction does your shadow point? Is this always true?

—DON STEINBERG

Twain Was A Child of The Comet

There has never been an American writer better known or better loved than Mark Twain. Famous for his novels, *Tom Sawyer* and *Huckleberry Finn*, and many other fine and funny works, he is also known for a lifelong connection to Halley's Comet.

Twain was born Samuel Langhorne Clemens on Nov. 30, 1835, in Hannibal, Mo. He died in Reading, Conn., on April 20, 1910. But what does this have to do with Halley's Comet?

It takes a year for the Earth to make one complete turn around the sun. It takes Halley's Comet 75 years to make one complete circle. So a Halley's Comet year is 75 Earth years. And Mark Twain was exactly one "comet year" old when he died. To be precise, he was one comet year, minus only 15 days.

When a comet is at the point in its orbit closest to the sun, we call it "perihelion." Halley's Comet reached perihelion on Nov. 16, 1835. Then it moved away into the outer reaches of the solar system for another 75 years. Its next perihelion was April 20, 1910. So Mark Twain was born when the comet was at perihelion, and died when it reached perihelion again.

Mark Twain lived one "comet year"

What Do They Mean by That?

A glossary of words used by astronomers and other scientists when they write about Halley's Comet and other celestial phenomena.

When you read about Halley's Comet, especially if you want to continue studying about it in other scientific works, you are likely to come across many unfamiliar words. This glossary should help you learn what those words mean:

Altitude: An angular measurement of position from the horizon upward. The zenith (the point directly overhead) has an altitude of 90 degrees from a level horizon.

Aphelion: The point in an object's orbit around the sun when it is farthest from the sun. (At this point, the object is traveling most slowly.)

Antitail: A trick of light that makes it appear that part of a comet's tail is pointing toward the sun. This sometimes happens when Earth crosses the comet's orbital plane.

Apparition: The period of time when a celestial object is visible from Earth.

Asteroid: A planet-like body in the solar system too small to be classified as a planet. Most asteroids orbit the Sun in a belt between Mars and Jupiter.

Astronomical Unit: A unit of measure in astronomy approximately equal to the average distance between Earth and the sun: 92,960,000 miles (149.6 million kilometers). Jupiter, for instance, is 5.2 astronomical units from the sun.

Azimuth: An angular measurement of position along the horizon, usually starting from north and moving clockwise. East is azimuth 90 degrees.

Celestial Equator: An imaginary circle in the sky formed by projecting the equator onto the sky. The celestial equator is 90 degrees from the celestial poles.

Celestial Poles: Imaginary points in the sky formed by projecting the Earth's north and south poles onto the starfield. The celestial equator lies midway (90 degrees) between the celestial poles.

Coma: The cloud of gas and dust surrounding the nucleus of a comet, which has not yet been swept into the tail by solar wind and solar radiation pressure. The material in the coma has been dislodged from the solid nucleus by solar energy when the comet moves close to the sun. The coma is seen as the comet's head.

Comet: A small object composed of frozen gases and dust in orbit around the sun.

Declination: A method of specifying the position of an object in the starfield. Declination in astronomy corresponds to latitude in geography — measuring positions north or south of the celestial equator in degrees.

Direct Revolution: The revolution of an object around the sun in the same direction as the planets. (Also called *prograde* or *posigrade* revolution.)

Direct Rotation: The spin of an object on its axis in the same direction as its orbital motion. (Also called *prograde* or *posigrade* rotation.)

Dust Tail: The tail portion of a comet composed of tiny solid particles. Solar radiation pressure pushes these particles of the coma away from the sun. The dust tail shines by reflecting sunlight and is yellowish-white. The dust tail is often noticeably curved.

Ecliptic: The plane of Earth's orbit around the sun, hence also the apparent path of the sun through the starfield (constellations of the Zodiac) in the course of a year as Earth revolves around the sun.

Ellipse: A closed geometrical figure formed when a plane (flat surface) cuts through a cone at an angle but does not cut through the base of the cone. Planets, asteroids and comets orbit the sun in ellipses.

Ephemeris: A listing that gives a celestial object's positions at various times.

Fluorescence: The process by which a gas emits light of certain wavelengths after absorbing light of different wavelengths. Comet gases fluoresce when the comet is close to the sun.

Gas Tail: See *ion tail*.

Gegenschein: Literally meaning "counterglow," this phenomenon of the zodiacal light comes from sunlight back-scattered from interplanetary dust located outside Earth's orbit and opposite the sun in the sky.

Giant Planets: Jupiter, Saturn, Uranus and Neptune.

Gravitation: The attraction of matter for other matter.

Head: The coma and nucleus of a comet.

Hydrogen Envelope: Seen only in ultraviolet light, this gigantic cloud of atomic hydrogen surrounds the comet's head.

Ion: An atom that has become electrically charged through the loss or gain of one or more electrons. Solar ultraviolet radiation is the principal reason neutral molecules become ionized in comets.

Ion Tail: The tail portion of a comet composed of ionized (positively charged) gases. The ion (or gas) tail is formed when particles in the solar wind interact with gases in the comet's coma. The ion tail shines by fluorescence and is usually straight and bluish-white.

Long-Period Comet: A comet with an orbital period greater than 200 years.

Magnitude: A measure of brightness in astronomy. The lower the magnitude number, the brighter the object. A star with apparent magnitude +1.0 is about 2½ times brighter than a star with apparent magnitude +2.0.

Meteor: The luminous streak in the sky when a particle burns up in the atmosphere (also called "a shooting star").

Meteorite: A natural particle that reaches Earth's surface from space after traveling through the atmosphere.

Meteoroid: A natural particle in space before it enters Earth's atmosphere.

Minor Planet: A planet-like body in the solar system too small to be classified as a planet. Most minor planets (or asteroids) orbit the Sun between Mars and Jupiter.

Nebula: A celestial cloud of gas and dust.

Node: A point along the orbit of an object where it crosses a reference plane. In the case of a comet, its ascending node is where it crosses the plane of Earth's orbit going north, and its descending node is where it crosses the plane of Earth's orbit going south.

Non-Gravitational Forces: Forces changing a cometary orbit that are not due to gravity, usually identified with rocket-like forces on the nucleus (the so-called "rocket effect.")

Nucleus: The central mass of a comet, composed of frozen gases (mostly water) and dust-sized particles of rock.

Oort Cloud: A spherical swarm of trillions of comets that surrounds our solar system. These comets orbit the Sun mostly at distances between 20,000 and 50,000 astronomical units.

Parallax: A means of calculating distance in astronomy by observing a nearby object from different positions to measure its displacement against a background of distant objects.

Parent Molecules: Water, carbon dioxide, hydrogen cyanide and other molecules containing carbon and sulfur that are believed to be the source molecules for many of the neutral and ionized atomic and molecular species observed in the coma and tail of a comet.

Perihelion: The point in an object's orbit around the sun when it is closest to the sun. (At this point, it is traveling fastest.)

Periodic Comet: An old name for a short-period comet (a comet that revolves around the sun in less than 200 years).

This sketch shows the component parts of a comet

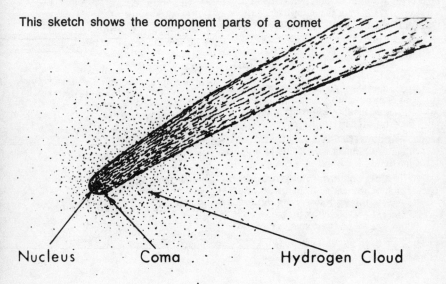

Nucleus   Coma   Hydrogen Cloud

Perturbation: Gravitational effects on the orbital motion of an object by masses other than the sun (usually major planets).

Planetoid: Another name for a minor planet (or asteroid).

Plasma: A "gas" of positive and negative ions.

Plasma Tail: Another name for a comet's ion tail.

Radiation Pressure: Electromagnetic radiation (light, infrared, X-rays, radio, ultraviolet, etc.) that can push materials away from the source of the radiation.

Revolution: The motion of one object around another. The Earth revolves around the Sun in one year.

Right Ascension: A method of specifying the position of an object in the starfield. Right ascension in astronomy corresponds to longitude in geography: measuring positions east or west along the celestial equator. Right ascension measures eastward from the vernal equinox, usually in hours (24 hours = 360 degrees; 1 hour = 15 degrees).

Rotation: The spin of a body on its axis. The Earth rotates once every 24 hours.

Scattering: Small particles (one micrometer to .1 millimeter in size) do not simply reflect light and make shadows but actually scatter the light that illuminates them in all directions. In some situations, forward-scattered light, appearing where a shadow would be expected, is actually brighter than back-scattered "reflected" light.

Shooting Star: Colloquial name for meteor.

Short-Period Comet: A comet that completes its orbit around the sun in less than 200 years.

Solar Wind: Fast-moving, charged subatomic particles that flow outward from the sun at speeds around 450 kilometers per second.

Striae: Narrow, rectilinear structures sometimes seen in a comet's dust tail. They are made of particles that were released at the same time from the nucleus and later disintegrate into fragments.

Sublimation: Changing directly from a solid state to a gaseous state without going through a liquid phase.

Synchrones: Particles released from a comet's nucleus at the same time. They are sometimes seen in the dust tail as straight or moderately curved structures.

Syndynames: Particles in the dust tail that are subjected to equal force.

Tail: Gases and solid particles from a comet's coma that are forced outward, away from the sun by the pressure of sunlight (dust tail) and solar wind (ion tail).

Zenith: The point in the sky that is directly overhead. The zenith will be different for observers in different locations.

Zodiacal Band: The faint glow seen along the ecliptic connecting the zodiacal light pyramids to the gegenschein.

Zodiacal Light: A general glow throughout the sky caused by sunlight scattered by interplanetary dust. It is brightest near the sun and along the ecliptic. The zodiacal light pyramids are often referred to as the zodiacal light.

Zodiacal Light Pyramid: A triangular glow seen on the western horizon after evening twilight and on the eastern horizon before morning twilight. It is the brightest component of the zodiacal light.

PORTRAITS OF A COMET

No, these are not actual photographs of Halley's Comet. They're very sophisticated artist's renderings of what the comet probably will look like when it gets closer. At the top is a drawing of the whole comet, streaking across the sky. Right is the artist's "close-up" of the comet's head. Clip these pictures and save them, and then compare them with the photographs that will be taken when the comet draws near.

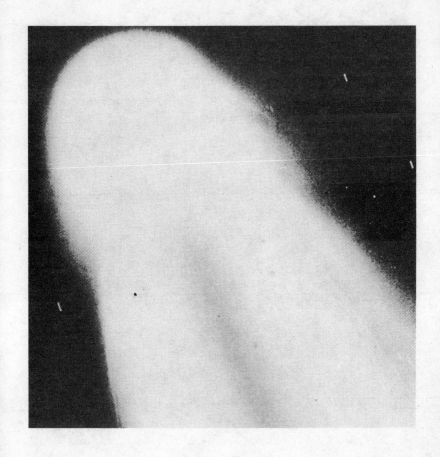

A Halley Album

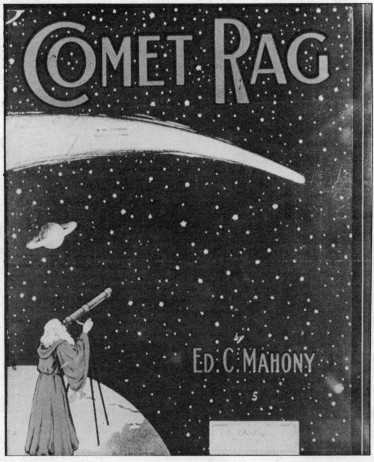

"Comet Rag" (shown here as a sheet music cover) was a popular hit of the day

This ad, showing a child sitting on the moon, features the comet

"Just trying to get a better look at the comet," the burglar tells the angry householder

This street scene in Punch shows a comet-watcher

Halley's Comet was as popular in 1910, the last time the comet was visible, as it is becoming today. Cartoons, fiction, advertising, even songs were dedicated to this celestial visitor. Here are some memorabilia of Comet Madness, 1910-style, courtesy of the Franklin Institute.

Comet Swift was captured on film in 1892!

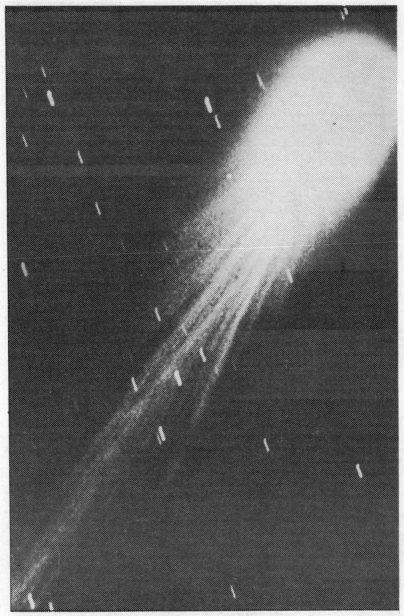

Comet Brooks was photographed in 1912

Comet West is a relatively new photo, taken in 1976

HALLEY'S COUSINS

Halley's is the most famous comet of them all, but it is by no means the only comet. We've had many such visitors over the years, and some have even been photographed. The white streaks in the photos resulted from leaving the shutter open — for up to several hours — to overcome the darkness.

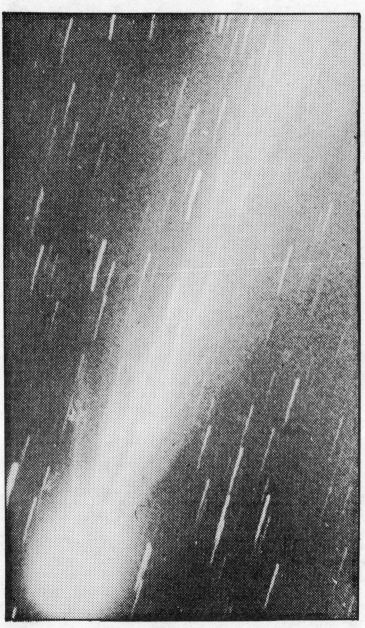
Comet Daniel: filmed in 1907

King Edward VII of England died in 1910

Crossword Solution

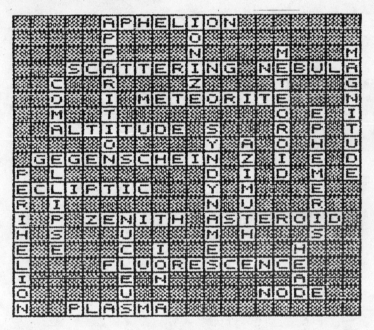